THE POETRY OF
MEITNERIUM

The Poetry of Meitnerium

Walter the Educator

Silent King Books

SILENT KING BOOKS

SKB

Copyright © 2024 by Walter the Educator

All rights reserved. No part of this book may be reproduced in any manner whatsoever without written permission except in the case of brief quotations embodied in critical articles and reviews.

First Printing, 2024

Disclaimer
This book is a literary work; poems are not about specific persons, locations, situations, and/or circumstances unless mentioned in a historical context. This book is for entertainment and informational purposes only. The author and publisher offer this information without warranties expressed or implied. No matter the grounds, neither the author nor the publisher will be accountable for any losses, injuries, or other damages caused by the reader's use of this book. The use of this book acknowledges an understanding and acceptance of this disclaimer.

"Earning a degree in chemistry changed my life!"
- Walter the Educator

dedicated to all the chemistry lovers, like myself, across the world

MEITNERIUM

In the cosmic dance of elements unseen,

MEITNERIUM

Where atoms whirl and molecules convene,

MEITNERIUM

There lies a realm of wondrous mystery,

MEITNERIUM

Where Meitnerium reigns in grand symphony.

MEITNERIUM

Born from the fusion's fiery embrace,

MEITNERIUM

In the heart of stars, it finds its place,

MEITNERIUM

A fleeting glimpse in the cosmic churn,

MEITNERIUM

A testament to the universe's yearn.

MEITNERIUM

With atomic number one-zero-nine,

MEITNERIUM

Meitnerium shines in the celestial design,

MEITNERIUM

A fleeting presence in the elemental array,

MEITNERIUM

Yet in its essence, it holds sway.

MEITNERIUM

In labs of science, it's carefully wrought,

MEITNERIUM

By skilled hands, its secrets sought,

MEITNERIUM

A fleeting moment in the laboratory's dance,

MEITNERIUM

As scientists delve into its enigmatic trance.

MEITNERIUM

Oh Meitnerium, element divine,

MEITNERIUM

In your nucleus, the secrets align,

MEITNERIUM

A realm of protons and neutrons held tight,

MEITNERIUM

In the quantum realm, where mysteries take flight.

MEITNERIUM

You dance with chaos, in the atomic storm,

MEITNERIUM

A fleeting whisper, yet your essence forms,

MEITNERIUM

In the tapestry of matter, you leave your mark,

MEITNERIUM

A testament to science's noble embark.

MEITNERIUM

In the annals of history, your name etched deep,

MEITNERIUM

A tribute to Lise Meitner's courage to keep,

MEITNERIUM

In the face of adversity, she dared to dream,

MEITNERIUM

And from her vision, Meitnerium did gleam.
MEITNERIUM

In the periodic table, you find your place,

MEITNERIUM

A symbol of discovery, a beacon of grace,

MEITNERIUM

For in your essence, we glimpse the sublime,

MEITNERIUM

The dance of atoms in the fabric of time.

MEITNERIUM

So let us marvel at Meitnerium's light,

MEITNERIUM

A fleeting wonder in the cosmic night,

MEITNERIUM

For in its existence, we find our quest,

MEITNERIUM

To unravel the mysteries, to boldly attest.

MEITNERIUM

To the beauty of nature's grand design,

MEITNERIUM

Where elements dance in the cosmic shrine,

MEITNERIUM

And in Meitnerium's fleeting embrace,

MEITNERIUM

We find a glimpse of the universe's grace.

MEITNERIUM

ABOUT THE CREATOR

Walter the Educator is one of the pseudonyms for Walter Anderson. Formally educated in Chemistry, Business, and Education, he is an educator, an author, a diverse entrepreneur, and he is the son of a disabled war veteran. "Walter the Educator" shares his time between educating and creating. He holds interests and owns several creative projects that entertain, enlighten, enhance, and educate, hoping to inspire and motivate you.

Follow, find new works, and stay up to date
with Walter the Educator™
at WaltertheEducator.com

www.ingramcontent.com/pod-product-compliance
Lightning Source LLC
LaVergne TN
LVHW010412070526
838199LV00064B/5273